1 MONTH OF
FREE
READING

at

www.ForgottenBooks.com

By purchasing this book you are eligible for one month membership to ForgottenBooks.com, giving you unlimited access to our entire collection of over 1,000,000 titles via our web site and mobile apps.

To claim your free month visit:

www.forgottenbooks.com/free1201293

ISBN 978-0-331-50120-9
PIBN 11201293

This book is a reproduction of an important historical work. Forgotten Books uses
state-of-the-art technology to digitally reconstruct the work, preserving the original format
whilst repairing imperfections present in the aged copy. In rare cases, an imperfection in
the original, such as a blemish or missing page, may be replicated in our edition. We do,
however, repair the vast majority of imperfections successfully; any imperfections that
remain are intentionally left to preserve the state of such historical works.

TWO LECTURES

ON

COMBUSTION:

SUPPLEMENTARY TO A

COURSE OF LECTURES

ON CHEMISTRY.

READ AT NASSAU-HALL.

CONTAINING

AN EXAMINATION

O F

Dr. PRIESTLEY's CONSIDERATIONS ON THE
DOCTRINE OF PHLOGISTON,

AND

THE DECOMPOSITION OF WATER.

BY JOHN MACLEAN,

PROFESSOR OF MATHEMATICS AND NATURAL PHILOSOPHY
IN THE COLLEGE OF NEW-JERSEY.

PHILADELPHIA:

PRINTED BY T. DOBSON, AT THE STONE-HOUSE, N° 41,
S. SECOND-STREET.

1797.

ADVERTISEMENT.

Owing to other engagements, a part only of the firſt of theſe lectures was read to the ſtudents—— They are now printed to ſave the young gentlemen the trouble of tranſcribing them.

J. M.

P. S. *It was not till after they were ſent to the preſs, that I was informed Mr. Adet had publiſhed an anſwer to Dr. Prieſtley's pamphlet.*

LECTURES

ON

COMBUSTION.

GENTLEMEN,

ALTHOUGH the confequences of the combination of different fubftances have been explained in the lectures which I have already had the honour of reading to you ; yet, as the appearances attending the expofure of inflammable bodies and metals to the air at high temperatures, are peculiarly ftriking, and have occafioned much difputation among philofophers, it will be proper to confider and compare the different opinions which have been held refpecting them.

Becher fuppofed that inflammable bodies contained a fubftance, which he called inflammable earth.

Stahl thought they included a peculiar principle ; to this he gave the name of Phlogifton. He fuppofed the phenomena of combuftion were owing to

its

its efcape, and the afhes or refidue of a burnt body was matter, by which the phlogifton had been confined. Thus, he thought, that in the combuftion of fulphur, the phlogifton was merely feparated from the fulphuric acid.

The difference among inflammable bodies, he attributed to the matter containing the phlogifton; and this opinion he confidered as confirmed, by the compound of potafh and fulphuric acid being readily converted into one of potafh and fulphur, by being mixed with powered charcoal, and raifed to a high temperature: for he fuppofed the charcoal communicated to the fulphuric acid the phlogifton which it had loft.

As all metallic bodies became converted into earthy-like fubftances when placed in certain circumftances, and as their original properties were reftored on their being mixed with inflammable bodies, and raifed to a high temperature, Stahl conceived that each metal confifted of phlogifton and a peculiar earth or calx.

It is eafy to perceive that Stahl did not attend to the influence the air of the atmofphere has on every ordinary procefs of combuftion. He was ignorant of the compofition of the atmofphere: he did not know that one of its parts combines with every inflammable body and metal which is burnt in it; that the new compound is in many cafes aërial, or gafeous, and in all exceeds in weight the combuf-

tible

tible fubftance; that this excefs of weight correfponds exactly to a like lofs fuftained by the atmofphere; that in fome cafes the whole, and in others
part, of the fubftance furnifhed by the atmofphere,
may be feparated with all its diftinguifhing properties; that in every inftance the whole of it may be
detached by the affiftance of other fubftances, which
have a difpofition to combine with it; and that in
proportion as the feparation takes place, the inflammable body, or metal, is recovered. In fine, it may
be perceived, that Stahl was unacquainted with
many of the phenomena, and formed a conjecture
to account for what he had obferved.

By the difcovery of feveral of the circumftances
juft mentioned, Mr. Lavoifier, a celebrated French
chemift, was induced to afcribe the change in the
properties of inflammable bodies, and metals, by
combuftion, to their union with the oxygen of the
atmofphere, and the great increafe of temperature
which attends fome of thefe combinations, to the
fame caufe that occafions a like phenomenon, during
the combination of other fubftances.

Mr. Lavoifier was foon fupported by feveral of
his countrymen, who were fatisfied there was no
occafion for fuppofing the exiftence of fuch a principle as phlogifton, to account for the phenomena.
Their doctrine has hence been called antiphlogiftic.

From its having arifen in confequence of the difcovery of feveral aerial or gafeous fubftances, it is

alfo

also named the Pneumatic Doctrine——It is the fyf-
tem in which you have been inftructed.

At firft it was ftrenuoufly oppofed by many philofophers, who, although they admitted Stahl's
theory was infufficient, yet thought the exiftence of
phlogifton was not incompatible with the difcoveries which had given rife to the antiphlogiftic doctrine, and was perhaps even neceffary for the explanation of fome, if not of all, of the phenomena.

In confequence, many attempts were made to re-
concile Stahl's notion with thefe difcoveries ; but
few phlogiftians agreed about the nature of phlogifton. Some of them confidered it as an immate-
rial principle ; others thought it was light ; others
hydrogen gas, &c. ; nay, it was not uncommon for
the fame perfon to waver between thefe different
opinions.

The event of the conteft has been highly honour-
able to the French chemifts. All philofophers now
living, except Doctor Prieftley, have acceded to
their doctrine.

The Doctor has often animadverted on their experiments and conclufions, and being perfuaded,
that what he has obferved has not been properly
attended to, or well underftood, he has lately publifhed a fmall pamphlet, entitled, Confiderations on
the Doctrine of Phlogifton, and the Decompofition
of Water. In it he has collected every thing he
confiders as material, whether as objections to the

pneumatic

pneumatic doctrine, or as arguments in favour of that to which he is attached. An examination of these, while it will unfold to you his modification of Stahl's theory, may serve to put you on your guard against falling into even that temporary delusion, which an erroneous opinion is so apt to produce, when supported by a celebrated name.

The work consists of a dedication, an introduction, and three sections: but as the discussion of the subject is confined to the sections, it will be necessary only to consider these.

In the first, he treats of the constitution of metals; in the second, of the decomposition of water; and in the third, of such objections to the pneumatic doctrine as could not conveniently be introduced into either of the other two.

Although this arrangement is not very well suited to our purpose, yet from an anxiety to avoid misrepresentation, I shall retain it; and for the same reason, instead of giving an abridged view of the Doctor's objections and arguments, I shall read his own words; but as he has not adopted the nomenclature in which you have been instructed, I shall mention the names that correspond with those which he employs.

In the section on the constitution of metals, after giving a very brief account of the principles of Stahl and Lavoisier, he says, " As a proof that me-
" tals are simple substances, and that they become
" calces

" calces (oxyds) merely by imbibing air, they al-
" lege the cafe of mercury, which becomes the
" calx called precipitate per fe (red oxyd of mer-
" cury) by expofure to the atmofphere in a certain
" degree of heat, and which becomes running mer-
" cury again by expofure to a greater degree of
" heat. They therefore think it impoffible not to
" conclude, that in all other cafes of calcination,
" as well as this, the only difference between the
" calx (oxyd) and the metal, is that the latter has
" parted with the air which it had imbibed."

This is certainly a very inaccurate ftatement. The
antiphlogiftians confider metals as fimple, becaufe
they have not difcovered them to be compound;
and they believe the fubftances, which the Doctor
calls calces, are compounds of metals and one part
of the air, which they call oxygen, becaufe they
cannot be formed without the prefence of oxygen;
they exceed in weight the metal, and this excefs,
correfponds to an equal lofs of oxygen; and the
oxygen may be recovered entirely, in the ftate of
gas, from the red oxyd of mercury, by fimply raifing its temperature; in part from the red oxyd of
lead, and black oxyd of manganefe by treating
them in the fame way; and may be wholly feparated from every metallic oxyd, by mixing them
with inflammable fubftances, and raifing the temperature of the mixture.

The

The formation and decompofition of the red oxyd of mercury is afcribed to the oxygen and mercury attracting each other, with different forces, at different temperatures; the partial decompofition of the oxyds of lead and manganefe is confidered as being owing to thefe metals retaining different proportions of oxygen with unequal force; and the effect of the inflammable fubftances is attributed to their having a ftronger difpofition, than metals have, to combine with oxygen: All which explanations are the more probable, from their correfponding with the laws of chemical combination, difcovered by obferving the action of different bodies on each other.

After giving the forementioned account of the foundation of the antiphlogiftic opinion, the Doctor attempts to fhew its infufficiency. " But this is " the cafe of only this particular calx (oxyd) of " this metal, and there is another calx of the fame " metal, viz. that which remains after expofing " turbith mineral to a red heat, which cannot be " completely revived by any degree of heat, but " may be revived in inflammable air (hydrogen " gas), which it imbibes, or when mixed with " charcoal, iron-filings, or other fubftances fnp- " pofed to contain phlogifton. And if this calx of " mercury, or (fuppofing it to contain fome acid " of vitriol,) [fulphuric acid,] this falt neceffarily " requires fome addition to conftitute it a metal,

B

" all

" all mercury muſt contain the ſame. For though
" with the ſame external appearance, the ſame
" metal may contain different proportions of any
" particular principle, as phlogiſton, they muſt be
" denominated different ſubſtances, if ſome ſpeci-
" mens contain this element and others be wholly
" deſtitute of it. All, therefore, that can be in-
" ferred from the experiment with the precipitate
" per ſe (red oxyd of mercury) is, that in this par-
" ticular caſe, the mercury in becoming that calx
" imbibed air, without parting with any, or very
" little of its phlogiſton ; and if we judge by the
" air expelled from the calces of metals and other
" circumſtances, there are few, if any of them, but
" contain more or leſs of phlogiſton."

From the following paſſages in Fourcroy's Elements of Chemiſtry &c. it appears the reaſoning in
the paragraph juſt read is founded on a miſtake.
Turbith mineral " when urged with a fire in a re-
" tort, at firſt becomes of a deeper colour, and is
" afterwards reduced to running mercury, giving
" out at the ſame time a conſiderable quantity of
" vital air (oxygen gas). Kunckel mentions this
" reduction. It ſucceeded with Meſſrs Monnet,
" Bucquet, and Lavoiſier, who traced it through
" all its circumſtances. I have repeated it ſeveral
" times with ſucceſs.

" Perhaps the reaſon why Mr. Baumé did not
" obtain running mercury, which has induced him

" to

" to affert that this yellow oxide does not refume
" a metallic form unlefs fome combuftible fubftance
" be added, was his not having applied to it a fuf-
" ficient heat."*

Thefe are confirmed by later obfervations,† and
they fhew, the mercury in turbith mineral, or any
fubftance into which it may be converted by a red
heat, does not require any addition to conftitute it
a metal.

It is true, the prefence of hydrogen gas, and
other inflammable fubftances, renders the reduction
eafier; but this it is contended, only proves, that
the fubftance is decompofed more readily, when
there is a body prefent, which has a difpofition to
combine with one of its conftituent parts, viz. the
oxygen, a circumftance analogous to many other
chemical decompofitions.

The antiphlogiftians are induced to believe the
hydrogen gas is not imbibed by the mercury, but
combines with the oxygen, becaufe, 1/. Oxygen
gas may be collected, when an oxyd of mercury
is reduced by fimply raifing its temperature: 2d.
When the reduction is performed in hydrogen gas
this difappears, no oxygen gas is obtained, but a
quantity of water may be collected: And, 3d. From
various experiments it feems, that water is a compound of hydrogen and oxygen.

* Fourcroy, Vol. II. p. 318. London Edition, 1790.
† Annales de Chimie, Tome 10me. p. 305.

That

That a higher temperature is neceſſary for the decompoſition of the ſubſtance from turbith mineral, than for that of the red oxyd of mercury, is not at all ſurpriſing ; they are different bodies, and muſt therefore be acted on by other agents, with different. forces.

The inference, which the Doctor draws from the experiments with the red oxyd of mercury, and from the gas obtained from the oxyds of other metals, cannot be admitted, until he proves the mercury and gas actually contain phlogiſton.

The Doctor proceeds—" I would obſerve in this " place, that it is aſſerted by ſome very able che- " miſts, that if the precipitate per ſe (red oxyd of " mercury) be made with proper attention, it will " be revived without yielding any air. This is alſo " the caſe with minium (red oxyd of lead) when " freſh · made. But this is owing, I doubt not, to " their wanting *water*, which I deem to be eſſential " to the conſtitution of every kind of air (gas) ; ſo " that they both contain the element of dephlogiſ- " ticated air (oxygen gas), though for want of " water, it is not able to aſſume that form."

I confeſs, I am unacquainted with the experiments, which have led to the aſſertion alluded to in this paragraph.—But, if water be eſſential to the conſtitution of every kind of air, and if mercury and lead are converted into their red oxyds, or calces, by uniting with the element of oxygen gas

or

or dephlogifticated air, what precaution can prevent the compounds from imbibing the water, if they. have a difpofition to do fo? The Doctor believes. they have fuch a difpofition; for except to thefe able chemifts, they have always afforded oxygen gas or dephlogifticated air, and to do this they muft, in his opinion, contain water.

The next paragraph is as follows. " That mer-
" cury may have the fame external appearance and
" all its effential properties, and yet contain differ-
" ent proportions of fomething that enters into it,
" is evident from the phenomena of its folution in
" the nitrous acid, and the revival of its calx in in-
" flammable air (hydrogen gas). According to the
" old theory, there is a lofs of fome part of its phlo-
" gifton in the folution of mercury in the nitrous
" acid, fince nitrous air (gas) is procured in the
" procefs: And though it may be revived from its
" precipitates (oxyds) by mere heat, yet if it be
" revived in a veffel of inflammable air (hydrogen
" gas), it will imbibe it in great quantities. Mer-
" cury revived in thefe circumftances muft contain
" more phlogifton than that which is revived from
" the fame calx by mere heat. But though mer-
" cury revived by mere heat after a folution in
" nitrous acid muft have a deficiency of phlogifton,
" and when it is revived from precipitate per fe
" (red oxyd of mercury) in inflammable air (hy-
" drogen gas) muft contain a redundancy of the
" fame

" fame principle, yet there will hardly be a doubt
" but that, in all chemical proceffes, it would ex-
" hibit the fame phenomena."

If two portions of matter exhibit the fame phenomena in all chemical operations, they ought moft
certainly to be confidered as the fame fubftance;
and as mercury revived from its oxyds by a mere
increafe of temperature, poffeffes all the properties
of that revived by the affiftance of hydrogen gas,
&c. it cannot be in any refpect deficient. It is in
vain to fay, that it muft be fo, becaufe " according
" to the old theory, there is a lofs of fome part of
" its phlogifton in the folution of mercury in the
" nitrous acid," and if it be revived in hydrogen
gas " it will imbibe it in great quantities;" it is
firft incumbent to prove, that the old theory, as
he calls it, is right, and that the gas which difappears, is actually abforbed by the mercury : but
thefe are juft the fubjects of difpute.

Having faid fo much on mercury he obferves.
" In all other cafes of the calcination (oxydation)
" of metals in air, which I have called the phlogifti-
" cation of the air, it is not only evident that they
" gain fomething which adds to their weight, but
" that they likewife part with fomething. The
" moft fimple of thefe proceffes is the expofing iron
" to the heat of a burning lens in confined air, in
" confequence of which the air is diminifhed and
" the iron becomes a calx (oxyd). But that there

" is

" is fomething emitted from the iron in this procefs
" is evident from the ftrong fmell which arifes from
" it. If the procefs be continued, inflammable air
" [hydrogen gas] will be produced, if there be
" any moifture at hand to form the bafis of it. From
" this it is at leaft probable, that, as the procefs
" went on in an uniform manner, the fame fub-
" ftance, viz. the bafis of inflammable air [hydrogen
" gas], was continually iffuing from it ; and this is
" the fubftance, or principle, to which we give the
" name of phlogifton."

 " That the effect of this procefs is not, as the
" antiphlogiftians affert, the mere feparation of the
" dephlogifticated from the phlogifticated air [the
" oxygen from the azotic gas] in that of the atmof-
" phere, I have proved in a courfe of experiments,
" in which I have fhewn that a confiderable part
" of the phlogifticated air [azotic gas] that is found
" after this procefs, is formed in the courfe of it,
" by the union of the phlogifton from the iron
" with the dephlogifticated air [oxygen gas]. And
" if the calcination of the iron in this procefs be
" always attended with the lofs of fome confti-
" tuent part of it, the fame is, no doubt, the cafe
" with all other calcinations of the fame metal, and
" alfo thofe of all other metals. And further, if
" the metals be compound fubftances, containing
" phlogifton united to fome bafe, the fame is the
" cafe with *fulphur* and *phofphorus*, becaufe they
       " become

" become acids when they are ufed in the fame
" procefs."

I do not know, that a fmell arifes from pure iron
when heated in air: but moft certainly there are
other metals, from which no fmell arifes, when
placed in the fame circumftances; and the Doctor
has faid, if the calcination of iron be attended with
the lofs of fome conftituent part of it, meaning
phlogifton, the fame is no doubt the cafe with thofe
of all other metals; and therefore even although
iron fhould afford a fmell, it is no proof of the efcape of phlogifton.

That a quantity of hydrogen gas is obtained
when the iron or air is moift is moft certain: but this,
it is infifted, is owing to the iron uniting with the
oxygen of the water, while the hydrogen affumes
the form of gas or air.

I have not been able to find any courfe of experiments, which *proves*, that azotic gas is formed
by expofing iron in confined air to the rays of the
fun concentrated by a burning lens: but Dr. Prieftley
informs us, that on examining the refiduum of fome
" pretty pure dephlogifticated air [oxygen gas]"
in which iron had been fired with affiftance of a
lens, " it did not appear that any phlogifticated
" air [azotic gas] had been produced in the pro-
" cefs,"* and the fame thing has been obferved by

* Experiments and Obfervations on Air, Vol. III. p. 481.

I   Lavoifier

Lavoifier,* confequently azotic gas cannot be form-ed by any thing from the iron uniting with oxygen, and the argument for the exiftence of phlogifton drawn from its fuppofed formation is invalid.

From the effects of air at high temperatures, the Doctor paffes to thofe of acids or metals. " Accor-
" ding to the antiphlogiftic theory, the inflamma-
" ble air [hydrogen gas] that is produced in the
" folution of metals in any acid, comes wholly from
" the water combined with it; and not at all from
" the metal diffolved. But the advocates for this
" theory do not feem to have attended to one ne-
" ceffary confequence of this fuppofition. Accor-
" ding to their own principles, water confifts of
" eighty feven parts of oxygen, to only thirteen of
" hydrogen in every hundred, which is nearly feven
" times as much of the former as of the latter.
" Confequently fince nothing but hydrogen efcapes
" in the procefs, there muft remain, from this de-
" compofition of the water, feven times as much
" oxygen in the folution. But both Mr. Lavoifier
" and Mr. de la Place fay [Examination of Mr.
" Kirwan's Treatife, p. 197, 198,] what I doubt

* Il eft de même extrêmement difficile d'obtenir du gaz oxigine parfaitement pur, il contient prefque toujours une petite portion de gaz azote, mais elle ne trouble en rien le réfultat de l'experience, & elle fe retrouve à la fin en même quantitè qu'au commencement. Annales de Chimie, Tome I. p. 26.

" not

" not is ftrictly true, that after the procefs the acid
" will faturate exactly the fame quantity (they do
" not fay more) of alkali, that it would have done
" before; whereas, with the addition of fo much
" oxygen, it ought to faturate confiderably more.
" If the oxygen from the decompofition of the wa-
" ter do not join that in the acid, what becomes
" of it?"

The anfwer to this queftion is very cafy—Far
from fuppofing the oxygen from the water joins
that in the acid, the antiphlogiftians believe it unites
with the metal, to enable this to combine with the
acid. They are of this belief, becaufe the metal is
precipitated in a ftate of oxyd when an alkali is
added to the folution; while the acid requires the
fame quantity of alkali to faturate it, and forms the
fame fubftances that it would have done before its
action on the metal.

Perhaps it was from being aware of this anfwer
that the Doctor has faid, in the next paragraph,
which concludes the fection, " If this cafe be ana-
" logous to that of the fuppofed decompofition of
" water by hot iron, the oxygen ought to be lodg-
" ed in the iron, and compofe *finery cinder* (black
" oxyd of iron). But this fubftance is not foluble
" in vitriolic (fulphuric) acid, if that be employed
" in the experiment; and when it is diffolved in the
" marine (muriatic) acid, it does not dcphlogifticate
" (oxygenate) it, as minium (red oxyd of lead)
" and

" and other fubftances containing oxygen, &c.
" It is evident, therefore, that there is no addition
" of oxygen in this procefs, confequently no de-
" compofition of water in the cafe, and that the in-
" flammable air (hydrogen gas) muft come from
" the decompofition of the iron."

It is rather furprifing the Doctor fhould affert, that finery cinder (black oxyd of iron) is not foluble in fulphuric acid, when, in page 505 of the third volume of his own experiments and obfervations, it is faid, that of fixty grains of finery cinder put into vitriolic (fulphuric) acid, * fifteen grains re-mained undiffolved. Befides, there is the moft fatis-factory evidence that iron, after its folution in the fulphuric acid, is in a ftate like that of the black oxyd or finery cinder. This fubftance diffolves in the acid without effervefcence, † and if the precipi-tate which the folution of iron in the fulphuric acid affords, on the addition of an alkali, be collected as foon as it falls, and dried in clofe veffels, it will be the black oxyd. ‡ There is no effervefcence during the folution, becaufe there is no hydrogen gas evolved ; and this, the antiphlogiftians believe, is owing to the iron being previoufly combined with enough of oxygen to fuperfede the neceffity of de-

---

* The quantity of acid is not fpecified.

† Lavoifier's Elements of Chemiftry, p. 88. Edinb. Edition, 1790.

‡ Fourcroy, Vol. II. p. 425.

compofing

compofing the water. The precipitate muft be dried immediately, and in clofe veffels, becaufe, from its minute divifion, it is very liable to be oxydated to a greater degree.

The other cirrumftance which he has adduced as a proof, that the black oxyd does not contain oxygen, is not more convincing. It certainly does not follow, becaufe muriatic acid can feparate a certain portion of oxygen from lead, when this is combined with a great quantity of that fubftance, that it fhould likewife feparate oxygen from iron, when this is united to a comparatively fmall quantity.

From this firft fection of the Doctor's work it feems, his objections to the opinion, that metals are fimple, and become converted into earthy-like bodies by an union with oxygen, are, 1ft. The fubftance which remains after expofing turbith mineral to a red heat, cannot be made to yield its mercury, unlefs it be mixed with bodies fuppofed to contain phlogifton. 2d. When iron is fired in air it emits a fmell, and if there be moifture at hand, inflammable air (hydrogen gas) is produced. 3d. Phlogifticated air (azotic gas) is formed during the combuftion of iron. 4th. If oxygen was feparated from hydrogen during the folution of iron in acids, thefe fhould acquire a proportional addition of ftrength. And, 5th. Iron diffolved in the fulphuric acid is not in the ftate of the black oxyd.

It alfo appears that the Doctor is of opinion,

1. That

1. That metals contain phlogiston.

2. That, in the formation of precipitate per se, (red oxyd of mercury) air is abforbed, and little or no phlogiston emitted.

3. That mercury revived from its calces (oxyds) by mere heat contains less phlogiston than when revived with the affiftance of inflammable air (hydrogen gas.)

4. That the fmell which arifes from heated iron, is owing to the phlogiston which is efcaping, and that the inflammable air (hydrogen gas) which is produced when moifture is at hand, is formed by the union of the phlogiston with the water.

5. That phlogifticated air (azotic gas) is formed by the union of phlogiston with dephlogifticated air (oxygen gas.)

6. That the inflammable air (hydrogen gas) emitted during the folution of iron in acids, is owing to the decompofition of the iron. And,

7. That *finery cinder* (black oxyd of iron) does not contain oxygen.

But from the facts mentioned in the review of the fection it follows, 1. The fubftance into which turbith mineral may be converted by a red heat, affords its mercury by a fimple increafe of temperature.

2. Mercury, revived from its oxyds by a fimple increafe of temperature, does not differ from that revived in hydrogen gas.

3. If

3. If a fmell arifes from heated iron, it is no proof of the emiffion of phlogifton.

4. Azotic gas cannot be formed by the union of oxygen with any thing emitted from heated iron.

And, 5. Iron, when diffolved in fulphuric acid, is reduced to a ftate like that of the black oxyd.

And further, if it can be proved that water is a compound of hydrogen and oxygen, and fufceptible of decompofition, as is infifted by the antiphlogif-tians, it muft alfo follow, that hydrogen gas, and other inflammable fubftances, affift the reduction of metallic oxyds, by combining with their oxygen. And that the hydrogen gas obtained by expofing heated iron to moifture, or by diffolving iron in di-luted fulphuric acid, proceeds from the decompofi-tion of the water, and not from that of the iron. —But the proofs of thefe will be exhibited in the examination of the fecond fection.

THE fecond fection of Dr. Prieftley's work is the moft important. He begins it with obferving, " The antiphlogiftic theory has received its greateft " fupport from the fuppofed difcovery, that water " is refolvable into two principles ; one that of oxy- " gen, the bafe of dephlogifticated air, and the " other, becaufe it has no other origin than water, " hydrogen, or that which, with the addition of " calorique, or the element of heat, conftitutes in- " flammable air."

The

The term hydrogen is derived from ύδωρ, aqua, and γεινομαι, gignor, and was designed to exprefs the principle engendering water. It has been criticifed by fome, who maintain it fignifies, engendered by water, and this is the fenfe in which Dr. Priefley feems inclined to underftand it. But, as has been well obferved by Mr. Lavoifier, it may be ufed in either of thefe acceptations; for where water is decompofed hydrogen is produced, and when hydrogen is combined with oxygen water is produced.

After giving an extract from a joint work by feveral of the antiphlogiftians, in which they declare their firm belief in the formation, the decompofition, and recompofition of water, the Doctor fays,

" Notwithftanding the confidence thus ftrongly " exprefled by thefe able and experienced chemifts, " I muft take the liberty to fay, that the experi- " ments to which they allude appear to me to be " very liable to exception, and that the doctrine of " phlogifton eafily accounts for all that they ob- " ferved.

" Their proof that water is decompofed, and re- " folved into two kinds of air, is, that when fteam " is made to pafs over red-hot iron inflammable air " is produced, and the iron acquires an addition of " weight, becoming what is called finery cinder; " but what they call oxide of iron, fuppofing that " there is lodged in it the oxygen which was one " of the conftituent parts of the water expended in

" the

" the procefs, while the other part, or the hydro-
" gen, with the addition of heat, affumed the form
" of inflammable air."

It ought to have been ftated, that in this experiment there is a lofs of water exactly equal to the
joint weight of the addition made to the iron, and
of the hydrogen gas obtained.

The antiphlogiftians fuppofe the addition made to
the iron to be oxygen, becaufe the compound re-
fembles, in every refpect, as the Doctor himfelf allows, that fubftance which is formed by burning
iron in oxygen gas, or in atmofpherical air ; and this
they confider as an oxyd, becaufe, 1ft. While it is
forming the oxygen gas difappears, and its weight
is exactly equal to that of the iron and the oxygen
confumed : *——And, 2d. When iron-filings are
mixed with red oxyd of mercury, and the mixture
made nearly red-hot, the iron is converted into the
fame black fubftance as in the laft experiment,
while the oxyd of mercury is reduced, and the
weight acquired by the iron correfponds to the excefs of that of the red oxyd above the mercury. †

---

* Quand on a donné à cette expérience toute l'attention
qu'elle mérite, l'air fe trouve diminué d'une quantité en poids
exactement égale à celle dont le fer eft augmentè. Annales
de Chimie, Tome I. p. 24.

† J'ai mêlè enfemble (fays Mr. Lavoifier), 450 grains
d'oxide rouge de mercure, par le feu, bien pur, & 100 grains
de limaille d'un fer tres doux, et qui n'etoit nullement atta_

The hydrogen gas is fuppofed to come from the water, and not from the iron, becaufe it is not ob_tained when the black oxyd is formed, without the affiftance of water; and, as will be fhewn, water may be formed by uniting oxygen with hydrogen.

The Doctor obferves, on the experiment with iron and water, " But in order to prove that this " addition of weight to the iron is really oxygen, " they ought to be able to exhibit it in the form of " dephlogifticated air (oxygen gas) or fome other " fubftance into which oxygen is allowed to enter, " and this they have not done. Iron that has really " imbibed air, or the common *ruft of iron*, has a " very different appearance from this finery cinder " (black oxyd of iron) being red, and not black; " and when treated in fimilar proceffes, exhibits " very different refults. Mr. Fourcroy fays (Exa_ " mination of Kirwan, p. 251.) that this finery cin_ " der is ' iron partially oxygenated.' But if that

qué de rouille. J'ai introduit ce mélange dans une petite cornue, & J'ai fait chauffer jufqu'au moment feulment où les vaiffeaux ont commencé obfcurément á rougir. Il ne s'eft dégagé aucun gaz pendant cette operation, fi ce n'eft une tres-médiocre quantité d'air fixe ou acide carbonique aériforme; elle n'excedoit pas deux ou trois pouces cubiques. Il a paffé dans la diftillation 415 grains de mercure coulant; ayant enfuite caffè la cornue, J'ai trouvé la limaille de fer dans l'état d'un fer brûlé, elle etoit friable, elle fe reduifoit aifément en poudre, elle etoit dans l'etat d'un veritable *ethiops* (black oxyd) & pefoit 132 grains. An. de Chimie, Tome I. p. 28.

" were

" were the cafe, it would go on to attract more ox-
" ygen, and in time become a proper ruft of iron,
" completely oxygenated. But this is fo far from
" being the cafe, that finery cinder never will ac-
" quire ruft ; which fhews, that the iron in this ftate
" is faturated with fome very different principle,
" which even excludes that which would have con-
" verted it into ruft."

The evidence which has been given, feems to me
to be fufficient to prove the addition made to the
iron muft be oxygen ; but more will be given in
the courfe of thefe lectures.

Without doubt common ruft of iron is very different from the black oxyd : but the Doctor is certainly miftaken in fuppofing this cannot acquire ruft ;
Mr. Fourcroy fays, it rufts fooner than common
iron, and every apothecary knows it does fo. Befides, we learn from the experiments of Meffrs.
Joffe and Fourcroy, that if ruft be made red hot in
a retort, a quantity of carbonic acid is difengaged
from it, and the iron remains in the ftate of black
oxyd. The ruft therefore is a carbonate of iron,
and muft contain all the principles which compofe
the black oxyd ; and this can contain none capable
of excluding that which would convert it into ruft.

The Doctor then remarks, " However, neither
" this, nor any other calx of iron, can be revived,
" unlefs it be heated in inflammable air (hydrogen
" gas), which it eagerly imbibes, or in contact with

" fome

" fome other fubftance which has been fuppofed to
" contain phlogifton. The probability therefore is,
" that the phlogifton then enters this calx of iron,
" replacing that which had been expelled to form
" the inflammable air. Nor can any inflammable air
" be procured in this procefs with fteam, but by
" means of fome fubftance which has been fuppofed
" to contain phlogifton. Where then is the cer-
" tain proof that water is decompofed in this pro-
" cefs ?"

The fuppofition that a body contains phlogifton
is no proof that it does fo.

" It may be faid, that the oxygen imbibed by
" this iron, being expelled by heat in contact with
" inflammable air (hydrogen gas), unites with that
" air, and with it conftitutes the water found af-
" ter the procefs. But for any thing that appears,
" this water may be that which the iron had im-
" bibed, and which can only be expelled from it
" by the entrance of that phlogifton which it had
" loft."

Doubtlefs you will be furprifed to hear, that water is the fubftance which the Doctor fuppofes is
capable, when combined with iron, of excluding that
which would convert it into ruft; and you will re-
collect that black oxyd of iron can be formed without the affiftance of water.

It is true, the Doctor obferves on the formation
of the black oxyd in oxygen gas and atmofpherical

air,

air, that " by far the greateſt part of the weight
" of dephlogiſticated air (oxygen gas) is water, and
" the air being decompoſed in the proceſs, the wa-
" ter is imbibed by the iron, and the acidifying
" principle (oxygen) contributes to form," according to the publication now under review, phlogiſticated air (azotic gas, *) but by his experiments and
obſervations, " fixed air (carbonic acid gas), with the
" phlogiſton which is at the ſame time expelled from
" the iron." †

It has been already ſhewn, that no azotic gas is
formed by the combuſtion of iron in oxygen gas ;
and the quantity of carbonic acid which has been
found in the remainder of oxygen gas, in which
iron has been burned, is very trifling, and is owing
partly to the gas containing ſome before the operation, and partly to plumbago contained in the
iron. ‡

But moreover, if the Doctor's explanation of the
formation of the carbonic acid gas be accurately examined, it will be found inconſiſtent with many of his
own principles.

He believes water is a conſtituent part of oxygen
gas, becauſe, from certain experiments, he has
inferred it conſtitutes one half of carbonic acid gas,

* See P. 15.
† Experiments and Obſervations, Vol. III. p. 551.
‡ Annales de Chimie, Tome III. p. 91—97.

and enters into the compofition of all aerial fluids : and from the quantity of water obtained on burning hydrogen in oxygen gas, he fuppofes it conftitutes nine parts in ten of oxygen gas. *

It will be fhewn in the next lecture, that his opinion refpecting the compofition of gafes is not well founded ; but for the fake of argument, let it for the prefent be granted, that oxygen gas is compounded as he fuppofes, and that carbonic acid gas confifts " of about one half water, and the other phlogif- " ton and dephlogifticated air † (oxygen gas) in the " proportion of one fourth of the former, to three " fourths of the latter." ‡

In page 159 of the firft volume of his Experiments and Obfervations, he fays, " In fix ounce meafures " and a half of dephlogifticated air (oxygen gas), " I melted turnings of malleable iron till there re- " mained only an ounce meafure and one third; " and of this twenty-feven thirtieths of an ounce " meafure was fixed air (carbonic acid gas)." Confequently the oxygen gas concerned in the formation of the carbonic acid gas, muft have occupied the fpace of 6,06 ounce meafures. §

* Experiments and Obfervations, Vol. III. p. 535.

† He means what he calls the acidifying principle or oxygen.

‡ Experiments and Obfervations, Vol. III. p. 536.

§ The volume of gas that difappeared was 5,16 ounce meafures, and that of the carbonic acid was ,9 of an ounce meafure.

An

An ounce meafure is equal to 1,8980 cubic inch, and therefore 6,06 ounce meafures are equal to 11,501 cubic inches.

A cubic inch of oxygen gas weighs ,34211 of a grain; * and 11,501 cubic inches weigh 3,9346 grains.

The weight of a cubic inch of carbonic acid gas is ,44108 of a grain; † and that of ,9 of an ounce meafure, or 1,7082 cubic inch is ,75345 of a grain.

The 3,9346 grains of oxygen gas, confumed in the experiment, confifted, according to the Doctor's eftimation, of 3,54114 grains of water, and ,39346 of a grain of oxygen: and the ,75345 of a grain of carbonic acid gas confifted of ,37672 of a grain of water, ,09418 of a grain of phlogifton, and ,28254 of a grain of oxygen. Therefore the oxygen admitted into the carbonic acid gas was lefs than that contained in the oxygen gas. What became of the reft? ‡

But the infufficiency of the Doctor's account of the formation of the carbonic acid gas will more clearly appear, on comparing the combuftion of iron with that of charcoal.

---

* Lavoifier's Elements, Edinb. edit. Appendix, p. 490.

† Ibid.

‡ The above calculation is made on the fuppofition that all the carbonic acid was formed during the procefs, although it is probable fome of it was contained in the oxygen gas before the combuftion of the iron.

" I heated

" I heated (fays he *) eight grains and a quarter
" of perfect charcoal, in 70 ounce meafures of de-
" phlogifticated air (oxygen gas) of the ftandard of
" 0,46, when it ftill continued 70 ounce meafures;
" but after wafhing in water, it was reduced to 40
" ounce meafures of the ftandard of 0,6, and the
" charcoal then weighed a grain and a quarter."

Suppofing this experiment to have been accurate,
which is not eafy to do, the quantity of oxygen gas
confumed was 19,47974 grains, and that of the car-
bonie acid obtained was 25,11509 grains. According to the Doctor, the firft confifted of 1,947974
grains of oxygen, and 17,531766 grains of water;
and the fecond of 9,418159 grains of oxygen,
3,139386 grains of phlogifton, and 12,557545
grains of water.

By this ftatement, the carbonic acid gas did not
contain fo much water as the oxygen gas; yet there
is none of that fluid depofited when perfect charcoal
is burnt in dry oxygen gas: the phlogifton in the
carbonic acid was not half the weight of the confumed charcoal; although the Doctor fays, page
166, Vol. I. of his Experiments and Obfervations,
that charcoal is very nearly pure phlogifton; the
oxygen in the carbonic acid gas was almoft five times
as much as that in the oxygen gas; and from the
experiment with iron, the oxygen in the oxygen

* Experiments and Obfervations, Vol. III. p. 377.

gas ought to have formed 3,73023 grains of carbonic acid gas, whereas the quantity faid to have
been obtained was 25,11509 grains.

It is plain, therefore, the formation of the carbonic acid gas cannot be accounted for, by fuppofing the two gafes to be compounded in the manner alleged by the Doctor, and charcoal to be pure
phlogifton.

But he has faid in page 547 of the third Volume
" that charcoal contains all the element of fixed air
" [carbonic acid gas], the acidifying principle as well
" as phlogifton."——However, even this fuppofition will not be fufficient to remove the difficulty.

It has been already mentioned, that of the
25,11509. grains of carbonic acid gas, 3,73023
grains fhould on the Doctor's principles be formed
by the affiftance of the oxygen gas : the phlogifton
fuppofed by him to be neceffary for this quantity
is equal to ,46627 of a grain, and ought to have
been furnifhed by the charcoal. The remainder
of the feven grains of charcoal fhould have confifted
of phlogifton and oxygen, in a proportion fit for
making, on the addition of water, the carbonic acid
gas. But the remainder is 6,53373 grains ; and as
water is fuppofed to conftitute one half of the carbonic acid gas, it ought to have formed 13,06746
grains, and thefe with the 3,73023 grains are lefs
than the quantity faid to have been formed by
8,31740 grains.

Now,

Now, as the formation of the carbonic acid gas by the combuftion of charcoal cannot be reconciled to the fuppofed compofition of the two gafes, and as the explanation which the Doctor has given, of the formation of the fame gas by the combuftion of iron is founded on the fame principles, it cannot poffibly be juft. And further, as his opinion of the compofition of finery cinder refts on the propriety of the explanation which he has given of the formation of the azotic gas, or of that of the carbonic acid gas, and as both have been fhewn to be wrong, it alfo muft be groundlefs.

In confirmation of his opinion the Doctor proceeds, " This is the more probable, fince when " any other fubftance, which is certainly known to " contain oxygen, is heated in the fame circum- " ftances, fixed air [carbonic acid] (which is al- " lowed to contain oxygen) is found, and this is " not the cafe with this calx of iron. If for ex- " ample precipitate per fe, or minium [the red " oxyds of mercury and lead] be heated in inflam- " mable air, the mercury and the lead will be re- " vived, and a confiderable quantity of fixed air [car- " bonic acid gas] will be produced at the fame time. " But if the air be previoufly expelled from the " minium which converts it into a yellow fubftance " called maflicot [yellow oxyd of lead] though the " lead will be revived, no fixed air [carbonic acid " gas] will be generated. Since therefore, the re-

E
" fult

" fult of treating finery cinder [black oxyd of iron]
" and maſſicot is precifely the fame, in the fame
" circumſtances, we are fully authoriſed to conclude
" that the fubſtances themfelves are fimilar, and
" confequently that the finery cinder contains no
" more oxygen than maſſicot."

Together with fome water, a fmall but not a
confiderable quantity of carbonic acid gas is com-
mouly obtained, when the red oxyds of mercury
and lead are revived in hydrogen gas. In one experiment made with red oxyd of mercury, fent to
the Doctor by Mr. Berthollet, he obtained 0,04
of an ounce meafure of carbonic acid gas; and in
another, made with red oxyd of lead, he procured
0,028 of an ounce meafure.*

Metallic oxyds are much difpofed to unite with
carbonic acid, and therefore the fmall quantity obtained in the above inſtances, might with fafety be
attributed, to their having attracted it from the
atmofphere and parted with it during the reduction.

But Mr. Berthollet informs us,† the oxyd which
he gave to the Doctor actually did contain carbonic
acid: On diſtilling 50 grains of it and receiving
the gas over lime water, although this was not at
firſt made turbid, after about a quarter of an hour
it depofited a confiderable precipitate. The precipitate he afcribed to carbonic acid uniting with the

* Experiments and Obfervations, Vol. I. p. 168.
† Annales de Chimie, Tome III. p. 91.

lime,

lime; and its flow formation to the acid being held in folution by a great quantity of oxygen gas, and to the portions of the carbonate of lime which were firft formed being diffolved by the unchanged lime water.

Dr. Prieftley obferves on this experiment of Mr. Berthollet, " The precipitate per fe with which " Mr. Berthollet furnifhed me, he fays, contained " a confiderable quantity of fixed air; and yet he " allows that when admitted to lime water, it did " not immediately make it turbid, which it is well " known a tenth part of the fixed air which I pro- " cured would have made it inftantly and complete- " ly white. The turbulency that came on after- " wards muft therefore have had fome other caufe, " probably fome acid of vitriol [fulphuric acid] in " the water of the trough in which the experiment " was made, and which gradually infinuating itfelf " into the lime water in his tube, would make *a* " *felenite* [fulphate of lime], a thing that has fre- " quently occurred in the courfe of my own experi- " ments, and which for fome time puzzled me not a " little."*

But although the carbonic acid gas which he ob- tained would make a certain quantity of lime water inftantly turbid, it is not well known that it would do fo when combined with feven or eight cubical

* Experiments and Obfervations, Vol. III. p. 559.

inches

inches of oxygen gas. Mr. Berthollet does not fay, that the water in the trough of his pneumato-che-mical apparatus was different from that in the veffel in which he received the oxygen gas; and I cannot believe a man of his accuracy would think of filling the trough, for fuch an experiment, with any other than lime water.

The yellow oxyd of lead may be formed by making the red oxyd red hot, or by expofing lead to the fame temperature and in contact with air. It affords no carbonic acid gas when it is reduced in hydrogen gas, not from its being deftitute of oxy-gen, but from its containing no carbonic acid. The temperature to which it has been previoufly fub-mitted is unfavourable to its union with carbonic acid; but if when cooled, it be expofed for a fhort time to the air, it will imbibe that fubftance and yield it when reduced in hydrogen gas.

It will be readily allowed by the antiphlogiftians, that if yellow oxyd of lead does not contain oxy-gen, the black oxyd of iron does not do fo either :; but the yellow oxyd cannot be made without the prefence of oxygen gas or atmofpherical air; it is heavier than the lead in its compofition; and its not affording carbonic acid is no proof that it does not contain oxygen.

The Doctor obferves further in fupport of his opinion, " In another important refpect finery " cinder [black oxyd of iron] and mafficot are fimi.

" lar,

" lar. They are both foluble in marine [muriatic]
" acid without dephlogifticating [oxygenating] it;
" which minium [red oxyd of lead] inftantly does.
" And yet Mr. Berthollet fays, (Annales de Chi-
" mie, Vol. III. p. 96,) that ' the heat by which
" minium becomes mafficot cannot change its na-
" ture.' What is the evidence of a change. in
." the nature of any thing, but a change of its pro-
" perties? On the whole, therefore, the proba-
" bility is, that when iron is converted into finery
" cinder, it lofes its phlogifton, and imbibes only
" water; and that when it is reconverted into
" iron, it parts with the water, and recovers its
" phlogifton. N. B. The experiment with the maf-
" ficot muft be tried prefently after it is made, fince
" it will very foon imbibe air from the atmo-
" fphere."

The red contains more oxygen than the yellow
oxyd of lead. Its parting with fome oxygen to the
muriatic acid is only a proof of what has been
already noticed, that the lead retains different
quantities of oxygen with unequal force.

The quotation from Mr. Berthollet is not exaft.
It is his intention to exprefs, that the carbonic acid
and azote, with which red oxyd of lead is com-
monly contaminated, may be feparated by a high
temperature, and yet the lead remain in the ftate
of an oxyd.

The

The reafon why the experiment with mafficot muft be tried prefently after it is made has been already affigned.

Now, confidering the manner in which the finery cinder and mafficot are formed, and that the different circumftances which have been alleged as proofs of their not containing oxygen are incompetent, the probability is, that they do not confift of water, and iron, and lead, deprived of phlogifton.

The Doctor then fays, " In this place I would " obferve that, if it be admitted that there is a prin- " ciple in inflammable air [hydrogen gas], which, " being imbibed by the calx of a metal, converts it " into a metallic fubftance, it will follow that the " fame principle is contained in charcoal, and other " combuftible fubftances ; becaufe they will all pro- " duce the fame effect, and therefore that the prin- " ciple of inflammability, or phlogifton, is the fame " in them all."

This will readily be admitted, but by the fame mode of reafoning it muft follow, that if a fubftance caufes the reduction of a metallic oxyd without communicating any thing to it, there can be no oc- cafion for the addition of any principle, and confe- quently no fuch principle as phlogifton exifting.

From the decompofition, the Doctor paffes to the recompofition of water : but the confideration of his obfervations on this fubject muft be deferred to the next lecture.

LECTURE II.

Dr. PRIESTLEY is as much diſſatisfied with the proofs of the recompoſition, as with thoſe of the decompoſition of water. " Another pre-
" tended proof [ſays he] that water is compoſed
" of dephlogiſticated and inflammable air [oxygen
" and hydrogen gaſes], is that when the latter is
" burned ſlowly in the former, they both diſappear,
" and a quantity of water is produced, equal to
" their weight. I do not, however, find that it
" was in more than a ſingle experiment that water
" ſo produced is ſaid to have been entirely free
" from acidity, though this experiment was on a
" large ſcale, not leſs than twelve ounces of water
" being procured. But the apparatus employed
" does not appear to me to admit of ſo much ac-
" curacy as the concluſion requires ; and there is
" too much of correction, allowance, and computa-
" tion, in deducing the reſult. Alſo it is, after
" all, acknowledged that, after decompoſing this
" quantity of the two kinds of air, and making all
" the allowance they could for the phlogiſticated
" air, or azote, in the dephlogiſticated air, they
" found fifty one cubic inches of this kind of air
" more

" more than they could well account for. This
" quantity therefore, and perhaps fomething more
" (fince the operators were interefted to make it as
" fmall as poffible) muft have been formed in the
" procefs. And when this kind of air, as well as
" inflammable [hydrogen gas] is decompofed to-
" gether with dephlogifticated air [oxygen gas],
" nitrous acid is produced. The probability there-
" fore is, that the acidifying principle, or oxygen,
" in the dephlogifticated air which they decom-
" pofed, was contained in that phlogifticated air
" [azotic gas], and that, had the procefs been con-
" ducted in any other manner, it would have affum-
" ed the form of nitrous acid. ; They acknowledge
" that, except when the inflammable air [hydrogen
" gas] was burned in the floweft manner, the
" water they produced had more or lefs of acidity."

The Doctor, at one time, believed that water
was compofed of oxygen and hydrogen. But as, on
repeating the experiment of burning hydrogen in
oxygen gas, he could not collect as much water as
was equal in weight to the gafes confumed ; and as
that, which he did obtain was mixed with nitrous
acid, he was induced to change his opinion, and to
fuppofe the water was not generated, but depofited
by the gafes during the combuftion, and that the
body formed was the nitrous acid.*

---

* Experiments and Obfervations, Vol. III. p. 43, et feq.

The

The antiphlogiftians alleged, when the experiment was performed on a large fcale, the deficiency
of water was very, trifling, and never more than
might with propriety be attributed to the unavoid-
able lofs to which fuch experiments were liable;
and they fuppofed the nitrous acid, found in the
water, proceeded from fome azotic gas having been
contained in the oxygen gas employed, and this the
more efpecially; as no way had been difcovered for
procuring oxygen gas perfectly free from azotic gas.

The experiment alluded to by the Doctor, in the
paragraph laft read, juftifies the reafoning of the antiphlogiftians. An account of it is to be found in the
eighth and ninth volumes of the Annales de Chimie.

The union of the two fubftances was effected
by filling a balloon with oxygen gas, adding to it
hydrogen gas in a fmall ftream, and fetting them
on fire by paffing the electric fpark through them.
To the balloon were connected refervoirs, called
gazometers, containing the two gafes. By certain
contrivances, thefe were made to fupply the balloon
with frefh portions of the gafes, as faft as the combination took place.

The experiment lafted 185 hours, and at the
end of it, there was collected of water 12 ounces
4 gros and 45 grains French weight, and there
remained in the balloon a quantity of gas.

By comparing the volume of the two gafes before the combuftion, with that of the gas remain-

F

ing in the balloon, and making every neceſſary correction for the difference of temperature and pref-
fure, it appeared, that 12 ounces 4 gros and
49,2270 grains of the gaſes had been confumed :
the difference between this weight and that of the
water is a mere trifle.

The water was perfectly pure : yet this was not
owing to the oxygen gas being free from admix-
ture ; for by a preliminary experiment it was difcovered, that 100 cubic inches of it contained three
of azotic gas ; and 467 French cubic inches of this
fubſtance were found in the balloon, at the end of
the experiment.

The cauſe of the purity of the water was difcovered by Mr. Seguin to be the flownefs with
which the combuſtion was conducted : for he has
afcertained, that, with materials of the fame kind,
the nitrous acid may be formed merely by carry-
ing on the combuſtion quickly, and by that means
raiſing the temperature to the point at which azotic
and oxygen gaſes act on each other.

This obfervation of Mr. Seguin has been confirmed by Meſſrs. Pelletier and Jacquin,* and alſo by
Mr. Van Marum ;† fo that pure water has been obtained m more than one experiment.

* Annales de Chimie, Tome X. p. 140.

† Ibid. Tome XII. p. 139. Mr. Van-Marum's words are,
" Dans une de mes expériences la combuſtion du gaz hydro-
" gène était très-lente, en employant trois heures et demie pour
" la confumption de mille pouces cubiqnes du gaz hydrogène,

But Dr. Prieſtley ſuppoſes the oxygen and phlogiſton formed azotic gas in Mr. Seguin's experiment. Certain it is, there were 51,7440 French
cubic inches of azotic gas found in the balloon at
the end of the experiment, above what had been
found in the oxygen gas before the combuſtion.

This quantity was ſuppoſed by Mr. Seguin to
have been owing to atmoſpherical air from which,
the gazometers could not be completely emptied,
before they were filled with the other gaſes: but
whether this be the true reaſon or not, for the appearance of theſe 51,7440 cubic inches, they could
not poſſibly have been formed in the manner ſuppoſed by Dr. Prieſtley.

The oxygen gas conſumed weighed 6209,869
grains and ought according to the Doctor to have
contained 620,9869 grains of oxygen or the acidifying principle; but 51,7440 French cubic inches of
azotic gas weigh only 22,9971 French grains. Nay
the whole quantity of azotic gas, found in the balloon, was equal only to 207,55348 French grains;
and yet, beſides oxygen, it ought to have contained
phlogiſton and water.

" & l'eau produite par cette expérience n'avoit abſolument
" point d'acide. Une autre fois la viteſſe avec laquelle l'air
" entroit dans la ballon étoit à peu près d'un tiers plus grande,
" & alors l'eau produite contenoit de l'acide foiblement ſen-
" ſible."

In

' In addition to thefe ftriking proofs of the incon-
fiftency of his principles, it may be remarked, the
phlogifton of the inflammable air [hydrogen gas]
ought to have weighed more than the whole quantity of azotic gas.

In page 290 of Vol. I. of his Experiments and
Obfervations, there is a calculation on the fuppofition that phlogifton compofes one half of hydrogen
gas: And in page 535 of Vol. III. he fays, " Wa-
" ter feems to conftitute about nine parts in ten of
" dephlogifticated air (oxygen gas), but there feems
" to be a much lefs proportion of it in inflammable
" air (hydrogen gas.)"

The hydrogen gas expended in Mr. Seguin's experiment amounted to 1039,358 grains; and if the
phlogifton be eftimated at only one fifth of that
weight, it will be 207,8716 grains, which is more
than the weight of the whole azotic gas.

Since, therefore, the azotic gas could not have
been formed by the oxygen and hydrogen, and fince
no other product was obtained than water, and
the weight of this correfponded to that of the two
gafes confumed, it may with fafety be inferred that
they formed water by their combination.

But, continues the Doctor, " The experiments
" which I made on the decompofition of thefe two
" kinds of air in *clofe veffels*, appear to me to be
" much lefs liable to exception, and the conclufion
" drawn

" drawn from them is the reverſe of that of the
" French philoſophers."

In what reſpect his experiments were leſs liable
to exception than thoſe of the French chemiſts, is
what I cannot comprehend. Theirs were performed on a very extenſive ſcale; great care was
taken to aſcertain the degree of purity of the gaſes
before combuſtion; and the apparatus was ſo con-
ſtructed that the reſults could be determined with
the greateſt nicety. The Doctor's, on the contrary,
were made with very trifling quantities of materials;
their purity was not tried; and their weight was
not accurately determined.

In one experiment, he employed ſuch a quantity
of the gaſes, as, he ſays, ought to have afforded a
grain of water; but he collected only a quarter
of a grain: in another, he ought to have got two
grains, whereas he obtained only a grain and an
half. And theſe are the experiments which he op-
poſes to, thoſe of the French chemiſts, and from
which he concludes, the water is not equal to the
weight of the gaſes conſumed! *

Satisfied, however, of the ſuperiority of his experiments, the Doctor proceeds to give their reſults.

" When dephlogiſticated and inflammable air
" (oxygen and hydrogen gaſes), in the proportion
" of a little more than one meaſure of the former

* Experiments and Obſervations, Vol. III. p. 45.

" to two of the latter, both fo pure as to contain no
" fenfible quantity of phlogifticated air (azotic gas),
" are inclofed in a glafs or copper veffel, and de-
" compofed by taking an electric fpark in it, a
" highly phlogifticated nitrous acid is inftantly pro-
" duced; and the purer the airs are, the ftronger
" is the acid found to be. If phlogifticated air (azo-
" tic gas) be purpofely introduced into this mixture
" of dephlogifticated and inflammable air (oxygen
" and hydrogen gafes), it is not affected by the
" procefs; though when there is a confiderable de-
" deficiency of inflammable air (hydrogen gas), the
" dephlogifticated air for want of it will unite with
" the phlogifticated air, and, as in Mr. Cavendifh's
" experiment, form the fame acid. But fince both
" the kinds of air, viz. the inflammable and the
" phlogifticated (hydrogen and azotic) contribute
" to form the fame acid, they muft contain the fame
" principle, viz. phlogifton."

" If there be a redundancy of inflammable air in
" this procefs, no acid will be produced, as in the
" great experiment of the French chemifts; but in
" the place of it there will be a quantity of phlo-
" gifticated air (azotic gas). A confiderable quan-
" tity of water is always produced in thefe decom-
" pofitions of air. But this circumftance only proves
" that the greateft part of the weight of all kinds
" of air is water. I have, in my experiments on
" terra ponderofa aerata (carbonate of barytes) de-
" monftrated

" monftrated that water conftitutes about half the
" weight of fixed air (carbonic acid gas.)"

It has been already fhewn, that hydrogen gas can
neither form azotic gas nor nitrous acid ; but it may
be worth while to point out the reafons for thefe
refults.

In the detail which is given of thefe experiments
in the third volume of his Experiments and Obfervations, there is no notice of any preliminary attempts to afcertain the degree of purity of the gafes ;
but it is there faid, the oxygen gas ufed in the two
trifling experiments formerly mentioned was got
from the oxyd of manganefe ; and that employed in
other experiments, during which the explofions were
performed in a copper veffel, was fometimes got
from the fame fource, and at others from the red
oxyd of mercury by the nitrous acid, and from the
red oxyd of lead.   Now it is well known, that all
thefe fubftances in general contain azote.—The firft
does fo, in fo remarkable a degree, that the firft portions of gas which it yields on being heated are
frequently pure azotic gas. The fecond, being made
with the nitrous acid, alfo contains fome of it. And
the third attracts it from the atmofphere, as the
Doctor himfelf has difcovered.

Mr. Cavendifh afcertained by his experiments,
that, if there be lefs hydrogen ufed than is neceffary
for the faturation of the oxygen, a quantity of nitrous acid is formed ; and that, if azotic gas be added

ded to the mixture, cæteris paribus, the quantity of acid is always increafed; but, that if there be a fuperabundance of hydrogen, no acid is produced.

Hence the reafon why the azotic gas was not affected in Dr. Prieftley's experiment, feems to have been, he ufed as much hydrogen as, with the azote contained in the oxygen gas and not attended to, was fufficient for the whole oxygen.

It has been already made evident, that water cannot enter into the compofition of oxygen and hydrogen gafes in the proportion alleged by the Doctor; and by a little attention to the experiments with the carbonate of barytes it will clearly appear, there is not the fmalleft foundation for the opinion that water is neceffary for the conftitution of carbonic acid gas.

The native carbonate of barytes was faid, by Dr. Withering and others, not to yield its carbonic acid at any temperature to which it could be expofed: but Dr. Prieftley found, that when the vapour of water was fent over it when red hot in an earthen tube, it afforded carbonic acid gas with the greateft rapidity, and in an equal quantity as when it was diffolved in the muriatic acid; while at the fame time fome of the water difappeared. He hence concluded, that water muft be a conftituent part of carbonic acid gas.

" Attending," fays he, " to the water expended " in the procefs, I found that I procured 330 ounce

I   " meafures

" meafures of fixed air (carbonic acid gas) with the
" lofs of 160 grains of water. According to this, as'
" the air weighed 294 grains, the water in the
" fixed air muft have been 80 parts of 147 of the
" whole.

" In another experiment, having previoufly found
" that three ounces of the terra ponderofa (carbo-
" nate of barytes) yielded about 250 ounce meafures
" of fixed air (carbonic acid gas), I attended only
" to the lofs of water in procuring it, and I found
" it to be one fifth of an ounce in two fucceffive
" trials. The quantity of fixed air (carbonic acid gas)
" would weigh 225 grains, and the water expend-
" ed about 100 grains, fo that in this experiment
" alfo the fixed air (carbonic acid gas) muft have
" contained about one half of its weight of water." *

This calculation, however, cannot be depended
upon ; for the lofs which the carbonate of barytes
fuftained was not examined, and the carbonic acid
gas muft have diffolved a quantity of water which
it would depofite on returning to the temperature
of the atmofphere.

To thefe objections, which were firft made by Mr.
Berthollet, the Doctor has returned, " I found very
" exactly how much fixed air [carbonic acid gas] a
" given quantity of this fubftance [carbonate of

* Experiments and Obfervations, Vol. I. p. 131.

q

" barytes]

" barytes] would yield by means of water, which
" appeared to be the very fame that it yielded by
" folution in fpirit of falt (muriatic acid), and that
" it yielded no air at all by mere heat without wa-
" ter. It was quite fufficient therefore to find how
" much water was expended in procuring any quan-
" tity of fixed air (carbonic acid gas) from this fub-
" ftance. And as there was no other fource of lofs
" of water befides the fixed air (carbonic acid gas),
" it could not but be concluded, that it entered into
" its compofition as a neceffary part of it, and in
" the proportion which I afcertained." *

In this anfwer, he has entirely overlooked the property which carbonic acid gas has of diffolving water. Every chemift knows it has that property,
and in a greater degree at a high than at a low
temperature. But the water is not neceffary to the
conftitution of the gas, becaufe it exifts before the
folution of the water ; and it may be deprived of the
water, by fulphuric acid or any deliquefcent fubftance, and ftill remain carbonic acid gas.

Befides, Dr. Hope, now Profeffor of Chemiftry in the Univerfity of Edinburgh, has difcovered that the carbonic acid can be feparated from
the barytes, by expofing the compound to fuch a
temperature as can be raifed in a fmith's forge. †

* Experiments and Obf rvations, Vol. III. p. 557.
† I dinburgh Philofophical Tranfactions.

To be fure, the difengagement of the carbonic acid takes place at a lower temperature when water is ufed; but this is only a proof that the feparation is promoted by the tendency which carbonic acid gas has to combine with water.

Hence then the celebrated experiments with the terra ponderofa aerata, or carbonate of barytes; afford no 'fupport to the Doctor's principles.

" The reafon, no doubt," fays the Doctor, in a note at the end of the pamphlet, " why, in the ex-
" periments of the French chemifts, the water they
" produced was not without acidity, whenever the
" flame they made ufe of was too ftrong, was that,
" in that cafe, more of the dephlogifticated air [ox-
" ygen-gas] in proportion to the inflammable [hy-
" drogen gas] was confumed, than when the flame
" was weak; fo that the refults of their experi-
" ments exactly coincide with thofe of mine."

When his experiments are accurately examined, they are found to confirm thofe of the French chemifts; but the reafoning in the note which has been read cannot be admitted.

The appearance of flame attends the combination of an inflammable fubftance and oxygen, when both are in the ftate of gas :—It is owing to their union taking place at many points at the fame time; but as the union depends as much on the one as on the other, a proportional quantity of each muft be as neceffary to exhibit a weak as a ftrong flame.

The

The Doctor further relates of his experiments,
" When the decompofition of dephlogifticated and
" inflammable air [oxygen and hydrogen gafes] is
" made in a glafs veffel, a peculiarly denfe vapour is
" formed, which the eye can eafily diftinguifh not
" to be mere vapour of water ; and if the juice of
" turnfole be put into the veffel, it immediately be-
" comes of a deep red, which fhews that it was an
" acid vapour.

" Since the acid that I procured in this procefs
" was in confiderable quantity, and no phlogifticated
" air [azotic gas] was prefent, (for in the laft of
" my experiments I did not even make ufe of an
" air pump, but firft filled the veffel with water, and
" then difplaced it by the mixture of the airs), I do
" not fee how it is poffible to account for the forma-
" tion of this acid but from the union of the two
" kinds of air ; and it can hardly be fuppofed that,
" in the very fame procefs, the decompofition of the
" fame fubftances fhould compofe others fo very
" different from each other as water and fpirit of
" nitre [nitrous acid]. I think I have fufficiently
" accounted for the refuit of the experiments made
" by the French chemifts on the common hypothe-
" fis, which fuppofes inflammable air to contain
" phlogifton ; but I do not fee how it is poffible for
" them to explain mine on theirs, according to which
" there is no fuch principle in nature. Upon the
" whole, it does not appear to me that the evi-
" dence, either for the compofition or the decompo-

" fition

" fition of water, is at all fatisfactory; and cer-
" tainly the arguments in fupport of an hypothefis
" fo extraordinary, and fo novel, ought to be of the
" moft conclufive kind."

Having in fome of his experiments emptied the
veffel in which the explofions were made of common
air, by means of an air pump, the Doctor fuppofed
it might be objected, that he could not entirely ex-
hauft the veffel; and it is on that account he has
mentioned, that in the laft of his experiments he did
not ufe an air pump : but the azote which occafioned
the production of the denfe acid vapour was contained in the oxygen gas which he employed.

The objections contained in this fection to the
conclufions drawn from the experiments which the
antiphlogiftians confider as proofs of the decompofition and recompofition of water, are,

1. Finery cinder [black oxyd of iron] does not
contain oxygen.

2. The weight of the water collected after burning inflammable and dephlogifticated airs [hydrogen
and oxygen gafes] is not equal to that of the airs
confumed. And,

3. Either the water fo obtained is mixed with
nitrous acid, or a quantity of phlogifticated air
[azotic gas] is formed.

The Doctor is of opinion,

1. That finery cinder [black oxyd of iron] confifts of water and iron deprived of phlogifton.

2. That

2. That when a metallic calx containing oxygen is reduced in inflammable air [hydrogen gas], fixed air [carbonic acid gas] is formed by the union of the oxygen with the phlogifton and water of the inflammable air. And,

3. That during the combuftion of inflammable in dephlogifticated air [hydrogen in oxygen gas], the phlogifton and oxygen form, according to circum-ftances, nitrous acid or phlogifticated air [azotic gas]; and the water obtained by the procefs is not generated, but from being a conftituent part of the two airs is depofited on their union.

But from what has been ftated in the review of this fection it appears,

1. The fame fubftance is formed by expofing iron to the fteam of water, by burning iron in oxy-gen gas, and by heating iron filings mixed with red oxyd of mercury.

2. Hydrogen gas is obtained when the iron is changed by being expofed to the vapour of water; but there is none afforded when the change is ef-fected by either of the other procefses.

3. The carbonic acid, which has been found after the reduction of certain metallic oxyds in hy-drogen gas, was previoufly contained in thefe oxyds.

4. The water which may be collected when hy-drogen is burned in oxygen gas is exactly equal in weight to the two fubftances which difappear.

4. The

4. The azotic gas which has been found in the refiduum was contained in the gafes before the combuftion.

6. The nitrous acid, formed when the combuftion was rapid, was owing to the union of azote with oxygen.

7. The experiments with terra ponderofa do not fhew that water is neceffary to the conftitution of the gafes. And confequently from thefe it follows,

That when water is brought in contact with red hot iron it is refolved into two fubftances, one of which combines with the iron, while the other affumes the form of gas—and, That water may be reproduced by reuniting thefe fubftances.

THE third fection begins as follows:

" Having confidered the evidence that has been
" alleged in fupport of the antiphlogiftic theory,
" and found it to be infufficient, I fhall in this fec-
" tion mention a few objections that may be made
" to it from other confiderations.

" 1. If inflammable air, or hydrogen, be nothing
" more than a component part of water, it could
" never be produced but in circumftances in which
" either water itfelf, or fomething into which wa-
" ter is known to enter, is prefent. But in my ex-
" periments on heating finery cinder [black oxyd
" of iron] with charcoal, inflammable air is pro-
" duced, though, according to the new theory, no
" water

" water is concerned. According to this theory,
" finery cinder, called the oxyd of iron, confifts of
" nothing befides iron and oxygen; and the char-
" coal, made with the greateft degree of heat that
" can be applied, is equally free from water; and
" yet when thefe two fubftances are mixed together,
" and expofed to heat, they yield inflammable air
" in the greateft abundance.

" This fact I cannot account for on the princi-
" ples of the new theory; but nothing is eafier on
" thofe of the old. For the finery cinder [the black
" oxyd] containing water, as one of its component
" parts, gives it out to any fubftance from which
" it can receive phlogifton in return. The water,
" therefore, from the finery cinder [black oxyd of
" iron] uniting with the charcoal makes the inflam-
" mable air [hydrogen gas], at the fame time that
" part of the phlogifton from the charcoal contri-
" butes to revive the iron. Inflammable air [hydro-
" gen gas] of the very fame kind is procured when
" fteam is made to pafs over red-hot charcoal."

Although hydrogen be a conftituent part of water it enters into the compofition of many other
bodies, and therefore the prefence of water is not
neceffary to account for its production.

The particulars of the experiment are related in
page 279 of the firft volume of the Doctor's Experiments and Obfervations. " Having" fays he,
" made the fcales of iron [black oxyd of iron], and

" alfo

" alfo the powder of charcoal very hot, previous to
" the experiment, fo that I was fatisfied that no air
" could be extracted from either of them feparately
" by any degree of heat; and having mixed them
" together while they were very hot, I put them
" into an earthen retort, glazed within and with-
" out, which was quite impervious to air. This I
" placed in a furnace, in which I could give it a
" very ftrong heat; and connected it with proper
" veffels to condenfe and collect the water which I
" expected to receive in the courfe of the procefs.
" But to my great furprife, not one particle of
" moifture came over, but a prodigious quantity of
" air, and the rapidity of its production aftonifhed
" me; fo that *I had no doubt* but that the weight
" of the air would have been equal to the lofs of
" weight both in the fcales [the black oxyd] and
" the charcoal; and when I examined the air which
" I repeatedly did, I found it to contain one tenth
" of fixed air [carbonic acid gas]; and the inflam-
" mable air which remained when the fixed air was
" feparated from it, was of a very remarkable kind,
" being quite as heavy as common air. The rea-
" fon of this was fufficiently apparent when it was
" decompofed by means of dephlogifticated air,
" [oxygen gas] for the greateft part of it was fixed
" air."

The Doctor now thinks this laft mentioned fixed
air was not a conftituent part of the inflammable

air,

air, but formed by the union of its phlogiston with the dephlogisticated air.

The reducing of wood to charcoal consists in separating the more volatile from the less volatile parts. This is done very imperfectly in common charcoal. The hydrogen especially is retained with so great force that the coal must be exposed to *an intense and long continued heat.*

It is not mentioned in the detail of the experiment, that the charcoal was previously exposed to the greatest degree of heat that could be applied, it is merely said it was made very hot, and that was very far from being sufficient.

Unglazed earthen vessels absorb moisture from the atmosphere very greedily, and it is scarcely possible to glaze accurately the inside of an earthen retort; at all events it is quite impossible, without breaking the retort, to know whether it has been perfectly done or not.

The charcoal, the iron scales (black oxyd), and the retort should all have been exposed separately to an intense and long continued heat, immediately before being used; the weight of the charcoal and iron before the experiment should have been compared with their weight after it; and the weight of the gases obtained instead of being guessed at, should have been accurately determined and compared with the loss sustained by the mixture. The experiment is, in its present state, of no value.

Mr.

Mr. Berthollet first objected to the experiment that hydrogen was with great difficulty separated from carbone, and the Doctor in reply to him has said in page 556 Vol. III. of his Experiments and Observations, " How obstinately charcoal retains
" water, is easily ascertained. For Mr. Berthollet
" himself would say, that when any particular de-
" gree of heat would not make charcoal yield any
" more inflammable air, there was no more water
" retained in it than the same degree of heat was
" able with its assistance to decompose. But by
" the assistance of finery cinder (black oxyd of iron)
" with even a much less degree of heat, it yields
" inflammable air very copiously, just as if steam
" had been made to pass over it in that heat; and
" judging from evident appearances, there can be
" no doubt but that with a sufficient quantity of
" finery cinder, to supply it with water, all the
" phlogiston in the charcoal, exclusive of that
" which contributed to the revival of the iron, will
" be converted into inflammable air."

From this answer, the Doctor seems to have mis-understood Mr. Berthollet. He has not said that charcoal retains *water* obstinately, his words are
" il paroît, par un grand nombre d'expériences que
" le charbon retient fortement de *l'hydrogène*; aussi
" avons nous distingué le principe charboneux au
" carbone du charbon ordinaire."

The

The refults of the Doctor's experiment may be accounted for in this manner. The carbonic acid gas was formed by the union of the oxygen in the oxyd of iron with the carbone in the charcoal; and the heavy inflammable gas' proceeded from the folution of fome carbone in hydrogen gas furnifhed either by the charcoal or moifture contained in the retort.

It can be no objection to this explanation, that the charcoal afforded an inflammable gas at a lower temperature when mixed with oxyd of iron than when ufed by itfelf. If the proportion of hydrogen be very fmall in comparifon with that of the carbone, the compound is folid at even a very high temperature; but when the proportion of hydrogen is greater, it is eafily made gafeous. In the foregoing experiment the proportion of carbone was diminifhed by the union of part of it with the oxygen of the oxyd of iron.

At all events the explanation offered by the Doctor cannot be a juft one. It has been already fhewn the fuppofition refpecting the compofition of oxygen gas is unfounded, and that, even after admitting that fuppofition, finery cinder cannot confift of water and iron deprived of phlogifton.

2. " Though the new theory, fays the Doctor, " difcards phlogifton, and in this refpect is more " fimple than the old, it admits another new prin- " ciple, to which its advocates give the name of " *Carbone*, which they define to be the fame thing
" with

" with charcoal free from earth, falts, and all
" other extraneous fubftances; and whereas we
" fay that fixed air confifts of inflammable air and
" dephlogifticated air or oxygen, they fay that it
" confifts of this carbone diffolved in dephlogif-
" ticated air, *fee Examination of Mr. Kirwan*, p. 79.
" Mr. Lavoifier fays, Ibid. p. 63, that ' wherever
" fixed air has been obtained, there is charcoal.'
" They therefore call it the carbonic acid.

" But in many of my experiments large quantities
" of fixed air have been procured where neither
" charcoal, nor any thing containing charcoal, was
" concerned, or none in quantity fufficient to ac-
" count for it. When the pureft malleable iron is
" heated in dephlogifticated air [oxygen gas] or in
" vitriolic acid air [fulphureous acid gas], a con-
" fiderable quantity of fixed air is formed. It is
" faid that *plumbago* is contained in iron. But it
" is not found in malleable iron, and leaft of all in
" the *air* that is expelled from it. Fixed air is
" alfo produced by reviving minium [red oxyd of
" lead] in inflammable air [hydrogen gas], and if
" charcoal of copper be heated in dephlogifticated
" air, a quantity of fixed air equal to nine tenths of
" the dephlogifticated air will be formed. More-
" than thirty ounce meafures of the pureft fixed-
" air were by this means procured from fix grains
" of this charcoal, which is made by the union of
" fpirit of wine and this metal.

" Laftly,

" Laftly, fixed air is procured in great abun-
" dance in animal refpiration. It is true that fixed
" air is procured by expofing lime water to atmo-
" fpherical air, but it is never procured by this
" means in air confined in any veffel. There muft,
" for this purpofe, be an open communication with
" the atmofphere. But fixed air will be procured
" in great abundance by breathing air contained in
" the fmalleft receiver, and efpecially if the air be
" dephlogifticated. It muft therefore be formed
" by phlogifton, or fomething emitted from the
" lungs, uniting with the dephlogifticated air which
" it meets there. It may be faid that fince we
" feed in a great meafure upon vegetables (and
" even animal food is originally formed from them)
" and this principle of *carbone* is found in all vege-
" tables, this may be the fubftance that is exhaled
" from the lungs. But fince in this procefs, it
" forms the fame fubftance that inflammable air
" from iron does with dephlogifticated air, or oxy-
" gen, it muft be the fame thing with it ; and then
" this *carbone* will only be another name for phlo-
" gifton."

The objection, that carbone is a hypothetical
being, was formerly made by Mr. Keir, and anfwer-
ed by Mr. Berthollet, " If there was no method,
" fays he, of procuring diftilled water, and that
" in the explanation of phenomena which are ow-
" ing to that fluid, it was confidered independently.

" of

" of the fmall quantity of falts which it holds in
" folution, would Mr. Keir look upon water as an
" hypothetical being of which no idea could be
" formed? Charcoal which has been well urged
" by the fire contains fometimes lefs than an hun-
" dredth part of foreign matter which has no in-
" fluence on its combinations; fometimes it contains
" much more: abftraction is made of that part
" foreign to its properties, and to avoid circumlo-
" cution, the name of carbone is given to the char-
" coal confidered in a ftate of purity."*

Notwithftanding what the Doctor has afferted, it
is fcarcely poffible to obtain iron free from plumbago; and this, from the quantity of carbone which
it contains, can, with a due proportion of oxygen,
make nearly four times its weight of carbonic acid.

The carbonic acid gas procured by the revival of
the red oxyd of lead has been already accounted for.

Charcoal of copper, as Dr. Prieftley calls it, is
made by paffing the vapour of alcohol or of oil of
turpentine through a red hot copper tube: a great
quantity of hydrogen gas is evolved and a black
fubftance collects in the tube. Of 446 grains of this
black fubftance obtained in one experiment, 28
grains were copper; and of 508 got by another, 19
grains were copper: the remainder when burned
afforded carbonic acid gas.

* Annales de Chimie, Tome X. p. 145.

Thefe

Thefe experiments prove that alcohol or fpirit of wine and oil of turpentine contain hydrogen and carbone; and that copper can feparate the carbone from the hydrogen. Charcoal of copper, therefore, is not a compound of fpirit of wine and copper, but of carbone and that metal.

That carbonic acid is formed during refpiration is moft certain, and that it is fo by the addition of fomething to the oxygen contained in the atmofphere is equally certain; but the Doctor has forgot when he fays, " it forms the fame fubftance which " inflammaable air from iron does with dephlogif- " ticated air, or oxygen."

In page 285 of the firft volume of his Experiments and Obfervations, when fpeaking of the carbonic acid gas obtained by burning the inflammable gas which is procured by paffing the vapour of water over red hot charcoal, he fays in the text. " That " the fixed air [carbonic acid gas] is not generated " in this procefs, is evident *from there being no fixed* " *air found after the explofion of dephlogifticated air* " *[oxygen gas] and inflammable air from iron.*" And in a note at the bottom of the page he obferves, " When I wrote this paper, I imagined that " the fixed air, which was found on the decompofi- " tion of this inflammable air with dephlogifticated " air, had been contained in the inflammable air. " But it will appear, that it muft have been formed " by the union of phlogifton [or inflammable air]

" and

" and dephlogisticated air, made by the explosion;
" though it is remarkable *that no fixed air is formed*
" *when the inflammable air from iron is used.*"

Besides in p. 562. Vol. III. he says of inflammable air from iron, " that it may not only be washed
" in lime water, but even be wholly decompofed by
" being fired together with dephlogifticated air,
" *without difcovering any fixed air at all.*"

Therefore the identity of carbone and the fup-
·pofed phlogifton has not been eftablifhed.

The third objection is a repetition of what he has
faid before. " 3d. The antiphlogiftians always fup-
" pofe azote, or phlogifticated air, to be a fimple fub-
" ftance, though I think abundant evidence has been
" given (and more will be found in my laft memoir,
" printed in the Tranfactions of the Philofophical
" Society at Philadelphia), that it is compofed of
" phlogifton and dephlogifticated air."

The abundant evidence which has been given
amounts to, an affertion that he has fhewn in a feries
of experiments that azotic gas is formed during the
oxydation of iron; and the circumftance of 51,744
cubic inches of azotic gas, having been found in the
refiduum of the great experiment made by Mr. Seguin and others, above what had been difcovered
in the oxygen gas before the combuftion.

It is fcarcely neceffary to remind you, the firft
is contradicted by his own experiments and thofe of
Mr. Lavoifier; and that the laft, cannot be account-

ed

ed for on his principles, even after granting a number of unfounded fuppofitions.

It is the Doctor's object, in the memoir which is to be publifhed in the fourth volume of the Tranfactions of the Philofophical Society of Philadelphia, to prove, that there is a greater quantity of oxygen in the atmofphere than is fuppofed by the antiphlogiftians, and was formerly believed by the Doctor himfelf; and that fome of the azotic gas found after the combuftion of certain fubftances in atmofpherical air, is formed by the union of their phlogifton with the oxygen of the atmofphere. But as the Doctor has not favoured us with a detail of his experiments, and as they bear the moft ftriking marks of not having been performed with accuracy, I will not take up your time with a review of them.

The Doctor then remarks,

" 4. As to the *new nomenclature*, adapted to the
" new theory, no objection would be made to it, if
" it were formed, as is pretended, upon a know-
" ledge of the real conftitution of natural fub-
" ftances; but we cannot adopt one, the principles
" of which we conceive not to be fufficiently afcer-
" tained. For other objections to this nomencla-
" ture, I refer to the Preface to Mr. Keir's excel-
" lent Dictionary of Chemiftry. However, whe-
" ther we approve of this new language or not, it
" is now fo generally adopted, that we are under
" the neceffity of learning, though not of ufing it."

Although

Although the new nomenclature is not ſtrictly
methodical, and its terms are rather uncouth and
harſh, yet as, in as far as the ſtate of our know-
ledge enables us to judge, it in general expreſſes either the properties or compoſition of bodies, I moſt
heartily recommend it.

The Doctor ſums up, " On the whole, I cannot
" help ſaying, that it appears to me not a little ex-
" traordinary, that a theory ſo new, and of ſuch
" importance, overturning every thing that was
" thought to be the beſt eſtabliſhed in chemiſtry,
" ſhould reſt on ſo very narrow and precarious a
" foundation, the experiments adduced in ſupport
" of it being not only ambiguous, or explicable on
" either hypotheſis, but exceedingly few. I think
" I have recited them all, and that on which the
" greateſt ſtreſs is laid, viz. that of the formation
" of water from the decompoſition of the two kinds
" of air, has not been ſufficiently repeated. In-
" deed, it requires ſo difficult and expenſive an ap-
" paratus, and ſo many precautions in the uſe of it,
" that the frequent repetition of the experiment can-
" not be expected; and in theſe circumſtances the
" practiſed experimenter cannot help ſuſpecting the
" accuracy of the reſult, and conſequently the cer-
" tainty of the concluſion.

" But I check myſelf. It does not become one
" of a minority, and eſpecially of ſo ſmall a mino-
" rity, to ſpeak or write with confidence; and

" though

" though I have endeavoured to keep my eyes open,
" and to be as attentive as I could to every thing
" that has been done in this bufinefs, I may have
" overlooked fome circumftances which have im-
" preffed the minds of others, and their fagacity is
" at leaft equal to mine.

" The phlogiftic theory is not without its difficul-
" ties. The chief of them is, that we are not able to
" afcertain the weight of phlogifton, or indeed that
" of the oxygenous principle. But neither do any
" of us pretend to have weighed *light*, or the ele-
" ment of *heat*, though we do not doubt but that
" they are properly *fubftances*, capable, by their ad-
" dition, or abftraction, of making great changes in
" the properties of bodies, and of being tranfmitted
" from one fubftance to another."

The experiments adduced in fupport of the antiphlogiftic doctrine are neither ambiguous, nor explicable on either *hypothefis*, nor few.—Thofe of the
French chemifts were performed with the greateft
care and nicety, and a few fuch are of more confequence than thoufands made without a due regard
to accuracy and precifion ; and if I miftake not it
has been fhewn, that they cannot be explained on
the Doctor's principles, and that his numerous experiments confirm theirs.

The experiment of the formation of water has
been frequently repeated. It has been performed
on a large fcale by Mr. Monge, by Meffrs. Lavoifier

and

Seguin, by Meſſrs Pelletier and Jacquin, and by Mr. Van Marum.

The Doctor deſerves credit for his candour; but I confeſs I am rather ſurpriſed, that after having calculated the weight of phlogiſton and oxygen, and ſupported a theory on the ſuppoſition of the calculations being right, he ſhould ſtate, as the chief difficulty attending that theory, the impoſſibility of aſcertaining the weight of theſe ſuppoſed ſubſtances.

However, if theſe ſubſtances or principles did actually exiſt, one or both of them would have weight; for charcoal, whether it be " very nearly pure phlo-" giſton," or contains " the acidifying principle as " well as phlogiſton," can be weighed.

It has been ſhewn, that the formation of carbonic acid gas by the burning of charcoal, cannot be explained on the Doctor's principles, if both the phlogiſton and the oxygen have weight; and it will be found equally inexplicable if one of them be deſtitute of that property.

Let it firſt be ſuppoſed the phlogiſton has no weight; and let it be again granted, that his experiment on the combuſtion of charcoal in oxygen gas was accurate. The oxygen in the conſumed charcoal could weigh no more than ſeven grains; and theſe, added to the tenth part of the weight of oxygen gas, would make 8,947974 grains. One half of the weight of carbonic acid gas is ſuppoſed to be water, and conſequently if the phlogiſton has no weight,

weight, the other half ought to be owing to the oxygen. But there could not be more of this oxygen
than 8,947974 grains, and these, with an equal
quantity of water, ought to have formed only
17,895948 grains of carbonic acid gas, while the
quantity said to have been obtained was 25,1150
grains.

On the other hand, if the oxygen be supposed to
be destitute of weight, and phlogiston to be heavy,
as the phlogiston could not exceed seven grains, the
quantity of carbonic acid gas should have been four-
teen grains.

Besides, the water in the oxygen gas ought, in
either case, to have exceeded that supposed to be
necessary for the constitution of the carbonic acid
gas. What became of this excess? Why did it not
combine with the one grain and quarter of uncon-
sumed charcoal?

Although it is more than probable that light, and
the cause which excites in us the sensation of heat
or caloric, are bodies; yet their existence as such
does not make a necessary part of the antiphlogistic
doctrine.

As the different parts of this section have no im-
mediate connection, it is unnecessary to make any
recapitulation.

The following note is subjoined to the last section: " N. B. For answers to the objections of
" Mr. Lavoisier and Mr. Berthollet to some expe-
" riments of mine relating to this subject, I refer
" to

" to the laſt edition of my Obſervations on Air,
" Vol. III. p. 554."

Such of theſe anſwers as were applicable to the
objections, which have been laid before you, have
been already conſidered.

I have now, Gentlemen, finiſhed the reading
and examination of Dr. Prieſtley's pamphlet. Perhaps, from having been ſo particular, I have almoſt
exhauſted your patience ; but I truſt you will ex-
cnſe me, as the fate of ſeveral important branches
of chemical ſcience is involved in that of this ſubject.

From the view which has been given of the different explanations of the phenomena of combuſtion it appears, that Becher's is incomplete ; Stahl's,
though ingenious, is defective ; the antiphlogiſtic is
ſimple, conſiſtent, and ſufficient ; while Dr. Prieſtley's, reſembling Stahl's but in name, is complica-
ted, contradictory, and inadequate. You doubtleſs
therefore will be inclined to prefer the antiphlogiſtic doctrine : Indeed you may adopt it with ſafety ;
for from being a plain relation of facts, it is founded on no ideal principle, on no creature of the ima-
gination ; it is propt by no vague ſuppoſition, by no
random conjecture ; it is dependent upon nothing
whoſe exiſtence cannot actually be demonſtrated ;
whoſe properties cannot be ſubmitted to the moſt
rigorous examination ; and whoſe quantity cannot
be determined by the teſts of weight and meaſure.

THE END.

Lightning Source UK Ltd.
Milton Keynes UK
UKHW022116081218
333475UK00006B/161/P